Breaking the Thought Barrier

Mimi Riser

Published by Mimi Riser, 2025.

BREAKING THE THOUGHT BARRIER

First edition. November 6, 2025.

ISBN: 979-8232029661

Written by Mimi Riser.

Table of Contents

The time has come, the Thinker said,
To talk of many things,
Of salami and space,
And cosmic force,
Of freedom and fiddle strings...

<><><>

BREAKING THE THOUGHT BARRIER

Rob & Mimi Riser

www.mimiriser.com

Smashwords Edition

<><>

TABLE OF CONTENTS

<><><>

"Pi"
Propellerheads-iconoclasts

It is time for all good propellerheads to come to the aid of their planet!

But what exactly *is* a propellerhead? Is it human? Or something cyborg? Does it speak the same language as everybody else? Can this species reproduce its own kind at will? *If allowed to function scientifically within intelligent surroundings, could a propellerhead ever win a Nobel Prize?...*

The title of Propellerhead applies here to any person engaged in scientific research without the support of a government grant, corporate sponsorship, a PhD from M.I.T., or a "full deck of cards." This latter credential being clearly nebulous, considering the colorful array of famous eccentrics who have lit up the history of science and mathematics.

Properly speaking, the propellerhead is a self-educated person who is drawn to science and math like a magnet. And the infinite number of unanswered universal questions simply won't let them go.

Possibly the most splendid example of propellerheadism were the Wright brothers. *Ahh*, yes, Wilbur and Orville—the W.O.W. brothers! Bicycle repairmen and designers. Self-educated scientists. In a mere seven years time these fellows achieved controlled motorized flight, something Leonardo de Vinci (1452-1519) and numerous other inventors after him had only been able to dream about.

Yes, Wilbur and Orville were indeed the *original propellerheads*: Sophisticatedly analytical, critically calculating, and forever researching the concept of propelled flight.

At one point, they discarded all the aeronautical information then available, deciding that it was both faulty and dangerous. In place of it, they designed and built a wind tunnel in which to analyze countless problems and situations concerning wings and wind. That wind tunnel proved to be their genie in a bottle. It was also the springboard into modern aviation.

The W.O.W. brothers began their exploits in 1896. They flew on December 17, 1903. This date, when you really think about it, should be a national holiday! And, according to accounts, they showed no trace of surprise when their flying machine actually worked; they were that sure of their calculations.

In much the same character, Albert Einsten, when his relativistic theory of gravitation was tested and verified in 1919, responded unemotionally, "But I knew that the theory is correct." When asked what if it had not been verified: "Then I would have been sorry for the dear Lord—the theory is correct." Einstein, by the way, taught himself mathematics, including calculus.

Granted, some might say, "Hey, that was then, and this is now!" But aren't we dealing today with the same universe they did a century ago? And since scientific theory is manmade, it is accepted or rejected by man. It is never absolute. Something rejected ninety years ago might well be regarded as priceless today. All knowledge, even if found under a cloverleaf, is always worth something—then and now.

The following pledge is intended to help each of us fully realize our freedom to explore the wonders of the universe with an unprejudiced and open mind, regardless of whether or not one's name is succeeded by half the alphabet.

The Propellerhead-iconoclastic Pledge

Stand vigourously! Face a mirror. While holding two spoons in your right hand, with the bowls extending in opposite directions (your propeller), raise that hand high and loudly proclaim:

I am proud to be a propellerhead! I am proud to be a propellerhead! I am proud to be a propellerhead! As an intelligent and insightful individual, I pledge to detect the rapture in relativity, to uncover the quintessential quanta in quantum physics, and possibly to discern why Bugs Bunny never won a Nobel Prize.

In the memory of Albert Einstein, modern science shall remain always "pleasing to the mind."

<><><>

Einstein Was Light...er, Right?

Several years ago, my wife and I were casually observing the west Texas night sky, when we spotted two low-flying colored lights on the eastern horizon. The green one appeared to be chasing the red one. After exchanging what looked like laser fire (honest!), the green one vanished.

The red one then banked and headed in our direction, accelerated, burst silently into a giant star-shaped light, and disappeared.

We would have yelled "Bravo!" and applauded, if we hadn't been so caught up in that *Star Trek* performance.

UFOs are everywhere lately, including the media. Our human fascination with these otherworldly wonders triggers many hard to answer questions, ranging from "Are they friendly?" and "Where do they come from?" to "Who designs those disks?" and "What kind of music do they like?"

My pet UFO question is "Have you seen the light?"

No, I'm not talking "Uncle John's Band" here, or even religion. I'm talking Einstein and 186,282 miles per second—the velocity of light. What we saw in the sky that night was a quickly accelerating, nonblinking, silent light, which burst into a magnificent "star" and poofed.

A year and a half earlier I'd witnessed the exact same thing (minus the galactic laser beam chase) while enjoying Perseids annual meteor show. A fast and steady light (not a meteor) accelerated toward me, discharged into a *Star Trek* blaze of light—*Wham! Outta here!*—and no sound whatsoever.

Now, let's analyze this...

Dr. Albert Einstein (1879-1955) who, in his own mind, went *where no man had gone before* or since, insisted that nothing is faster than the velocity of light, nothing in the universe can exceed lightspeed.

The reasoning behind this statement is that mass (some large or small clump of matter) collects more mass (that clump becoming heavier) as it approaches the speed of light. Thus, by the time it reaches 186,282 miles per second, it is simply too massive (i.e. dense) to *jump the fence* and exceed that velocity. In other words, the collected extra baggage is holding things back, preventing that clump from the jump.

To draw a sports analogy, consider that if a baseball pitcher *could* throw a ball that fast, and if the batter *could* make contact with it, the weight (solid mass) of the ball, when it reached the batter, would completely shatter the wooden bat (to say nothing of the poor batter).

Einstein's insistence that nothing can exceed the velocity of light gets tested regularly in those massive electromagnetic merry-go-rounds known as particle accelerators. These bombastic devices, with the aid of super-conducting magnets, spin subatomic particles up to a cat's whisker shy of 186,282 miles per second, never jumping over that, never getting a speeding ticket.

Time and time again, like a band of Buddhist monks, physicists hear themselves chanting, "Einstein was right...Einstein was right...Einstein was right..."

And yet, in spite of all this, I still see that silent, star-shaped burst of *Star Trek* light...

So, we have this now well-worn and time-after-time tested theory that nothin' can beat Lightspeed. He runs too fast and wears a big "C" on his chest (the math symbol for "velocity of light").

But now let's suppose, hypothetically, that somewhere and at sometime some scientist and/or mathematical wizard comes up with one of those "They'll never believe it!" ideas. This idea centers on a device for some flying craft that instantly transforms the excess mass collected at the jump-the-fence point into *Energy*. Thus, we solve two problems at the same time: You get rid of the excess baggage and—*Bam!*—you also get the instant, extra energy that lets you and your flying craft make the jump into *warpspeed*.

Hypothetical, to be sure. Plausible enough for hard science-fiction. But it *is* based on history's most famous equation: $E=mc^2$. (Although it belongs to Uncle Albert, the "E" represents Energy, not Einstein.)

Simply, the equation states that Energy (E) equals Mass (m) times the Velocity of Light squared (c^2).

What it also means is that Energy and Mass are interchangeable; under certain conditions energy can become mass, and vice versa. And so, back in our hypothetical science-fiction wizard scene, his strange device would *zap* that extra mass-baggage into energy, and, in turn, jump his craft past the flashy character with the big C on his tights.

Mathematically, and in real life, this Mass-into-Energy theory was proven true in 1945 in New Mexico, when Robert Oppenheimer, Enrico Fermi, and a family of other young hot-brains put together and popped off the world's first atomic bomb.

And, if you're over fifty years of age, you probably remember Chuck Yeager and those other flying aces of the 1950s seemingly forever cracking the sound barrier (as well as a few windows).

Breaking the sound barrier creates tremendous noise (big surprise) and shock waves. Nuclear (fission) explosions do the same, along with their spherical waves of light. When you break a barrier of any kind, you are creating violent disturbances, the reactions of which must, in part, include a transformation of the mass and energy properties.

But these UFOs are silent. And that *Star Trek* blaze of light is a silent flash.

Perhaps, in this situation, we're dealing with *implosion*, the opposite of explosion; everything gets yanked inward instead of being blown outward. Implosion is an intensely complicated subject. But this simple explanation gets across the idea that, when UFOs accelerate and jump-the-fence into *warpspeed*, they may just be doing so via some aspect of implosion. (At any rate, like a housefly, they seem to exhibit complete freedom of motion. They also seem to contribute nothing to our global problem of noise pollution.)

We've all seen those famous photos of these vehicles—they resemble a discus. When the brawny discus thrower spins and lets that disk fly, it sails in aerodynamic perfection, then slaps the ground. But no problem. The discus is designed to absorb any and all shocks without the slightest damage to itself. No sphere exists that can make the same claim. Spheres, of all types, always bulge on impact, eventually loosing their shape. The discus is both completely shockproof and virtually indestructible.

Speaking hypothetically once again, could we assume that what Chuck Yeager experienced each time he broke the sound barrier must also be experienced when breaking the *light barrier*? Violent aggregations of turbulent universal forces, created by the give-and-take of energy and mass, are more than willing to rip to shreds any ill-designed vehicle. The only qualified shape that comes to mind is the discus.

All in all, it is rather amazing how little we actually know about UFOs, in light of how far our technology has come since 1945. In that year the boys at Los Alamos passed the first "bomb test," and (if you can somehow manage to ignore its devastating and brutally destructive powers) it did open new doorways in science. But some *cosmic doorways* may have been opened as well. In 1947, UFOs were popping up everywhere; they became a very hot commodity in the media.

During these past sixty-plus years, our rapidly developing technology has cluttered the planet with stadium-sized radar antennas, by which even a flock of geese can be spotted and analyzed. But, apparently, no UFOs can be seen?

Some conspiracy advocates claim, and not without justification, that the major world powers already know more than they're willing to admit about UFOs. If this is the case, then they also know who's actually "behind the wheels" of these *fly-by-night* craft (which also means, the pilots probably all have day-jobs).

This is only speculation, of course. In fact, after hundreds or more documented UFO sightings by thousands of people (some of them even reputable) all we *do* have *is* speculation. We can't even be sure if these shiny pyrotechnical disks are from Mars, the dark side of the Moon, or Mount Etna, let alone who's flying them.

To the masses, UFOs are marvelous entertainment. To modern science, they sit as a royal pain in the process of observation and analysis. As an example: astronomers require the convenience of cosmic regularity, knowing, for instance, that the Perseids meteor shower will be performing its mid-August show on schedule. From that knowledge, they can then work comfortably with data they can rely on, and get things done.

None of this is possible with the elusive UFOs. In physics, concepts must be validated by mathematics. In mathematics, concepts need to be applied to the physical world. Where the UFO is concerned, neither of these approaches have enough hands-on information for anyone to work with.

Science is filled with speculations—they're called Theories. But these theories generally have at least a few logical justifications backing them up.

UFOs haven't allowed us to get close enough to formulate any real theories, let alone justify them.

So, it looks like our chant will simply have to continue for the time being... *Einstein was right... Nothing can exceed the velocity of light!*

On the other hand, if something ever did, even as we stood there watching it...

How would we be able to tell?

<><><>

Don't Let the Gravity Get You Down

The world of science has two problems. One is gravity. The other problem is that gravity is not recognized as a problem.

Gravity is considered a fact of life, one of those you-can't-do-anything-about-it things. But suppose we take a look at it, and think a little about what we're actually seeing.

Sir Issac Newton, in his *Principia* (1687), declared that any two bodies, large or small, throughout the physical universe, attract each other with a force that varies "directly as the product of their masses and inversely as the square of the distance between them."

Hey, folks, *any* two bodies, throughout the physical universe, attracting each other? This ain't gravity. But what is it?

Science sees gravity as earth-bound, and gravitation as universal. Um, excuse me, but aren't we earthlings part of the overall universe?

Also, Albert Einstein, the guru of twentieth century science, in his generalized field theory, presented the theoretical relationships involved in gravitational and electromagnetic fields as being *the same.*

So, we do have forces at work here. But the problem is that we have *forces* (plural), namely electromagnetic and gravitational.

Electromagnetism is normally seen as separate from gravity. When you talk one, you usually are not talking the other. But Einstein's field theory appears to co-relate them, suggesting that Newton's gravity actually possesses electromagnetic properties.

The question now quickly arises, what exactly are we talking about here?

The answer is gravity. The problem is gravity. The *real* question, therefore, must be "What *is* gravity?"

The dictionary describes this force as "the downward acceleration of terrestrial bodies due to the gravitation of the earth, modified by centrifugal force, etc."

Well, that's a mouthful. No wonder some people think science is hard. That definition really didn't tell us much, other than what even the scholars aren't really sure about.

What is needed is a close and clear look at this universal force; otherwise we, as a society, will never ever get past this situation. As great a thinker as he was, Newton was still a man of his own time. As long as we continue to view gravity as Newton did, we'll forever have one technical foot in the Middle Ages.

Einstein? Relativity, you ask? OK, sure, he greatly redesigned the mathematical concepts of this universal force. But he still called it *gravity*.

Alfred Korzybski's classic book on general semantics, titled *Science and Sanity*, clearly explains the power of The Word. It's like a TV game show; you guess the right word and you go home with a bundle of cash. In essence, *the word is the meaning*, so an entire culture can easily be influenced by, and swept up into what a word *happens* to mean.

So maybe we should drop that word? (*Bang! Crash! Ah-h-h-h*, there's yet another crack in the foundation of physics). By not using that word we begin deprogramming our three-hundred-year-old association with it.

Let's do some thinking. In our universe we have these natural forces: electricity and magnetism (collectively known as electromagnetism), gr-gr-gr...(*that* force), space and time (collectively known as spacetime). There's also the king-of-the-hill, Light, but that sits on its Constant Velocity Throne by itself.

It seems that in physics, which is the study of matter (mass)-in-motion, whenever you have, in a given situation, a force that you can't *plug in* (i.e. electromagnetism), then it's gotta be...that force.

Why is it that on this planet, out here in the cosmos, whether we're talking hypothetically or realistically, if you can't "plug it in," it's gravity? (Oops!)

The situation in physics has become one of prejudiced assumptions, a very dangerous approach to the natural physical world, especially because that world is constantly fluctuating between waves and particles. When you're looking for real answers you have to look at what is there in front of you, see what is *given*, analyze it objectively, and proceed from there. So, because physics largely deals with vortices, spheres, curves, and wavy lines, let's look at something that, in many ways, represents the four aforementioned shapes: Centrifugal Force.

Most people associate *that word* (you know the one I mean) with centrifugal force, even though what we have here is primarily inertia, which is the tendency for all things to continue their motion undisturbed. It seems we also have *momentum*, that quantity of motion that is equivalent to the object's mass and velocity. We can actually relate a lot of physics to centrifugal force, since nearly all of the motions and actions of the four universal forces are essentially curved in nature.

Now let's have a little fun and jump feet first into the deep end of the pool, so to speak. There is this *creature* known as spacetime curvature. No, it's not a new ride at Disney World. *This* is Einstein's explanation of how gr...gr...(*that force*) is created. It's a dandy, so strap yourself in; here goes...

When light, traveling through the universe at its never changing velocity (speed) encounters any object (mass), it gets bent (warped) a bit. As does the object. Also getting warped are space and time (along with Einstein's sophisticated sense of humor).

Anyhow, what you have here is mass (made up of electromagnetism), space and time—three of the four universal forces—being bent, shoved and crammed together in one small space. Kinda like big city rush-hour traffic, only more interesting.

Now, folks, once again, let's play this word game, pretending we don't know what physicists claim this creates. But, to be sure, *something* gets conjured up here, manifested, and it is powerful.

Light travels at the speed of 186,282 miles per second (never slowing down to take in the scenery). We have all this stuff here packed together like a roll of salami. Now what?

Pause for a moment, remembering the information on centrifugal force and all that stuff about vortices, spheres, curves and wavy lines representing physics. Whenever you bend a straight line and get a curve, the natural inclination for that curve is to complete itself and form a circle. If the circle is three-dimensional, it will form a sphere.

OK, back to the action!

We have this salami spacetime curvature, all crammed tightly together, and it's really powerful. It wants to do something. Oh, yeah, curves want to form circles; better yet, spheres. Let's form a big sphere. Let's cover the earth with this power! Salami Power!

Sounds like the return of the Roman Empire or something. Makes me think of Ben Hur racing 'round and around Rome in that wild chariot race.

But wait, what power *are* we referring to here?

The world of science generally agrees that our entire world is electromagnetic in nature. Electrons and magnetism, a duo from which everything is given birth.

You can have magnetism isolated and on its own, but wherever you have the electric, you also have the magnetic. Magnetism is a natural property of electricity, but not vice versa.

It is magnetism that keeps the electrons close at home in orbit around the atom's nucleus, just as it is the force behind electricity. Thus, magnetism forms the disciplined focus for the electron.

As it is in the microcosm, so it is in the macrocosm. Since our solar system has long been stamped with the tag of being a big—no, *gigantic*—copy of the atom's structure, what is the force keeping our nine planets in orbit around their nucleus...er, I mean Sun? If the atom is *enclosed* by magnetism, and the atom is our universe in micro, why is it supposedly you-know-what that is keeping our solar bodies in orbit?

If you have a length of straight copper wire and *plug it in*, you get electromagnetism. The magnetic force circles the copper wire. If you form a circle with the wire, and plug it in, the magnetic force circles in and out of the wire loop, perpendicularly.

Now, apply this electromagnetism to our space-warped sphere. Can you see it? We're talking *magnetism* here, not the g-word, as being the predominating force in the universe, both on this planet as well as throughout the cosmos.

Circles, spheres, vortices, and wavy lines dominate physics, as do spacetime, gravity, and electromagnetism. Did I say gravity?... Uh-huh. Shall we consider it an *aspect* of magnetism, an activator of impulses in certain instances of motion? Shall we combine the two together and do what all of science loves to do, give it a name? How about *Gravito-magnetism*?

The Romans would've enjoyed that. Sounds like the name, maybe, of some prominent member of the Roman Senate.

And what, if anything, did the Romans call the predominate force in the universe?

Whoever constructed Stonehenge, did they have a name for this force?

When the Egyptians built their still mystifying Pyramids, did they, perhaps, somehow call upon an overriding universal force to help them construct those monolithic temples?

Could Stonehenge and the Pyramids have *been* built if those cultures *had* a view of something akin to gravity?

How a society views its world and environment has a powerful impact on its day-to-day activities. And our concept of gravity was born during the late Middle Ages, when the Inquisition was still staring over everyone's shoulder. If we take a fresh look at magnetism with a very objective mind, we may see it as it really is.

Also, aspects of gravity might take on a new light, revealing physical properties of both that are really much the same. This collection of fresh information could even offer some possible clues toward solving the Rubik's Cube of physics, Einstein's unified field theory.

<><><>

Tai Chi and the Science of Bowing a Fiddle

"When the bowing gets tough, the tough get bowing."

The scene: You're a double bassist in your final year of music conservatory. You take, as a recreational elective, a one-semester Tai Chi martial arts course. *Bang!* Your approach to using a bow suddenly evolves, and neither you nor your bow technique are ever the same again.

The story: My third year of college studies at the New School of Music (1982-83) in Philadelphia, Pennsylvania was intense. My attitude was "I'm here in college and studying with these great musicians, so I gotta get the most out of it!"

I did. Unfortunately, the body, mind, and spirit had gotten stuck in attack mode, and I couldn't find the relaxation button. Lots of stress, something I haven't always dealt well with.

My body had been injured, then healed that spring. Now the mind and spirit needed healing. Some type of change was being called for. In June, my principal teacher (a bassist with the Philly Orchestra) transferred me to Temple University's College of Music.

I desperately needed to find that damned relaxation button.

Somebody at Temple threw me into a large "winds and brass ensemble" in which I was the only string player. The conductor told me I could play anything on the page that happened to be in bass clef. He placed me behind the winds, at center, with brass on both sides of me. (Having studied Buddhism, I knew I was trying to find *Center*, but this was ridiculous.) His other comment was, "Don't worry about anything, Rob, just play loud!"

Just play loud. This meant applying as much weight and arm pressure as possible to the bow and on the strings. Somewhere in my conscious mind some specific concepts began taking shape and getting analyzed.

Three and a half months of marching band music... I exhaled! My mind smiled and proclaimed "Hey, Buddha, I found that relaxation button. I found Center. Thank you!"

And I laughed. Every Monday, Wednesday and Friday afternoon, after the usual three-hour orchestra rehearsal, I'd have a two-hour winds and brass rehearsal, and laugh some more.

Fourteen weeks of fun and music, and nothing but fun and music! Went through a cake of rosin real fast. We played three concerts, one at Temple and the other two at the annual Teachers Convention in Pittsburgh.

Did I mention that the majority of the students in that winds and brass ensemble made up the Temple University Band, featured at all the home football and basketball games?

Party people.

At this same wondrous time, I was taking a Tai Chi martial arts course as a free-elective. Three mornings a week, Monday, Wednesday and Friday. And as a result, Tuesdays and Thursdays became rather boring.

In this marvelous course we were taught an interweaving series of slow movements involving the feet, legs, torso, head, arms and hands. It was ancient oriental dance, and it always slapped a smile on my face.

I practiced these moves every night for an hour or two. Then one night a sphere of powerful energy—unseen but clearly felt—suddenly bloomed between my arms, heart charka, and hands. Doing what I considered to be the logical thing, I ignored it and continued through the series of movements. The energy soon diminished.

In the next Tai Chi class I detailed my odd experience to the instructor, who agreed that I did the right thing, saying: "If it happens again, don't focus on it; just see it through and continue moving through the cycle of gestures." Somehow, though, I did notice that this energy topic made him a bit nervous...

And looking back on it, maybe he had a valid reason to be nervous. In Tai Chi you have a series of ancient oriental martial arts movements, which for some people (like me) cause a tapping of *cosmic energy*. These movements, if performed in a specific manner, utilize and integrate the etheric vibrational force that's the heartbeat of all situations, paranormal and otherwise, in the universe. Strong stuff.

The instructor had suggested not focusing on it. But what would happen if you just utilized the *sense* of that energy?

During the 1983-84 holiday season, I considered whether or not this concept could be applied to bow technique. String players are taught early to "get into the string." The pre-eminent technique for getting the student to play "in" the string is an acoustic monster known by the name of Longtones. It's a mercilessly slow dragging of the bow across each of the strings just above the bridge, and with as much muscle tension and pressure as you can find. This sound closely resembles fingernails dragged slowly across a chalkboard.

But it works. And it also allows the player to get more time and volume out of each bow stroke; especially important for basses since they have the shortest bows in the business.

Now, I thought, what if the player found some other source for that muscle pressure? What if the player simply let the *weight* of the action itself pull the bow? (The way one Tai Chi movement flows out of and into the next.) Would any of that "in the string" value be lost? Would the resulting sound still be deeply resonant, projective, and ultimately as professional as the famous Philadelphia Sound?

So I spent the holidays applying this concept to bass solos and common audition excerpts. I was in no hurry. I wanted only to test this bow concept thoroughly and over all areas of the instrument.

You tune your violin, viola, cello, or bass to the standard A-440 frequency (A-441 in Boston); you now learn to tune your bow to *Tai Chi energy*, cosmic rhythm and flow and follow-through.

It is a simple feel, adapted from the practice of the movements. Once I got this feel, I could easily shift back and forth between the two bowing techniques whenever I chose to. It's really that simple. (I just preferred the Tai Chi technique feel and sound.) And by the way, concerning double bass, this bowing technique serves both the French and German bows.

After New Year's, I brought it out to meet the public. My stand partner (I'll call him Ted) in a weekly professional-level sight-reading orchestra that I played in noticed a sound difference and asked about it.

"New strings, Rob?"

"No, Ted, new bow technique."

Schooled in Philadelphia, Ted was a professional musician all the way around, and had some intellectual leanings as well. I briefly explained what I had discovered and how I was utilizing it. Ted didn't say too much in response other than he really liked the sound. But even more, he didn't criticize the unusual source of the sound and its concept.

That new sound was amazingly deep, "from way inside the bass," and clean. The famous bass solo from Mahler's first symphony sounded nearly like a human voice.

The three hour reading session that night provided a great test to see just how practical and functional this new bow approach would be under normal conditions. It was, in a word, magnificent.

Any concerns and reservations I might have had were now wiped clean from the slate.

When college resumed, I had my weekly bass lessons with my Philly Orchestra teacher. When I mentioned my new bow technique and the Tai Chi relation, he lost it.

Yelling "No!" he leaped out of his chair, grabbed my bass and bow, and spent several manic minutes exclaiming: "You have to *force* the bow into the string with lots of muscle tension, like this. Your arm should feel tired and exhausted after a rehearsal or performance!"

I didn't say anything, but I was thinking "man, that's dumb." It's also very sexual, very *guy*. Plus it was very "papa" patriarchy. And since all fulltime professional double bassists play with heavy bows, ideal for head-whacking potential muggers after dark, I had additional problems with those remarks. (Related thought: If an Olympic runner competed with too much muscle tension, it could cost that runner both the race and his or her career.)

However, he was "big time." I was a comparative nobody. Understanding how, in any profession, everybody knows everybody, I kept my mouth shut, told no one else, and continued playing and developing this fascinating technique on my own.

Thus, the situation now became a game, on the level of "You can see and hear what I'm doing, but you obviously will never know exactly *what* I'm doing and thinking." Very cool. A chess player's delight.

As soon as I realized the yin-yang implication here, bass playing took on a whole new dimension. Most fiddle players (in classical music, at least) *attack* the instrument, insisting upon *pulling* the sound *they* want from it, via the bow. Another Philadelphia Orchestra bassist I knew even went so far as to say: "You have to dominate the bass, Rob; show it who's boss."

I thought differently. Were we playing football here, or acoustical musical instruments?

Every fiddle (violin, viola, cello, and double bass), of every quality, has hidden within itself an individual natural sound, as well as an individual personality and character. The Tai Chi approach to playing allows the instrument to project its own unique voice, one that will always be beautiful, if you take the time to listen, if you allow it to happen.

On a spiritual level, if you want to get the most out of Nature, work *with* Her; don't rape Her. She is not a prisoner of war to be tortured for Her secrets.

And like that, this bowing technique is a combination of two functions. One, the matriarchal respect for Nature, and letting a fiddle's natural voice sing. Two, the Zen Buddhist concept of getting out of your own way, and helping *you* let your fiddle sing.

One final technical point: This bow technique is not a replacement for longtones. Longtones get you there "in the string." The Tai Chi "go with the flow" technique keeps you there, but without the unnecessary wear and tear on both you and the instrument.

As an example, when you're driving a four-speed car, your first three speeds get you to highway velocity. Once you've reached that velocity, your fourth gear allows you and your car to relax and travel along without burning the amount of gas your third gear would. Creative intelligence provides such things.

It took me a while to complete the formation of this bowing technique. Once I did, though, the sound I produced was a very rich and projective voice, both crisp and clear. Elegant.

A case in point: In 1989, the Harrisburg (Pennsylvania) Symphony Orchestra was rehearsing Bruckner's fourth. I was on the second stand. The two fellows on the first stand kept looking my way and talking to each other. At rehearsal break they asked if they could try out my bass. Eventually I realized that my sound had grabbed their attention. No, they couldn't duplicate it. Yes, I continued keeping mum about the creation of that elegant sound.

To be honest, by then my inner guidance system was already losing patience with playing papa-patriarchy games. It was time to begin playing my own game.

"The true value of a human being can be found in the degree to which he has attained liberation."

~Albert Einstein

<><><>

The Selfless Ones

Like large sponges with heads, arms and legs, many people are self-absorbed, almost completely focused on their own affairs and thoughts. If you should ever ask them why they're so spongy, they'd probably throw you any one of a number of mind-your-own-business retorts (if not actually throwing you a punch and telling you to keep it).

A much safer maneuver is to open any dictionary and discover the several pages devoted to SELF words. If these words are quickly analyzed, three general groups emerge:

Group 1 – "I am the world!" – These include self-deluded, self-opinionated, self-abuse, self-willed, self-pride...(yes, there are quite a few)...self-conscious, self-conceit, self-love, and selfish, among others. Ask one of these people for the time of day, and they'll probably tell you, "I get off of work at four o'clock."

Group 2 – "I can do anything better than you!" – This collection of high-gear go-getters includes self-determination, self-sufficient, self-heal, self-improvement, self-contained, self-adjusting...(there will be a quiz on these later)...self-made, self-esteem, self-guidance, self-starter, self-help, self-acting, self-government... I could go on, but you get the idea, right? There's more meat to this group—although they do remind me of some professional musicians I have known ("I can play Mendelssohn better than you!"). Ask them for the time, and you'll likely hear, "Why aren't you wearing a watch?"

And now we come to the "How did you happen to meet Harvey, Mr. Dowd?" – Group 3 – Self-forgetful, self-immolation, self-imposed, self-perception, self-respect, self-luminous...(unfortunately, there aren't enough of these for a quiz)...self-control, self-sacrifice, selflessness... Ask these free-flowing folks for the time of day, and you'd likely hear, "Well, it was three o'clock a couple of hours ago." If there's one descriptive term for this group, it's *pleasant*. Makes me envision poet Walt Whitman sitting aglow in a grassy field. Transcendental is another term. They do seem a bit beyond the here and now.

How about a blend of the aforementioned groups? Oh, forget Group 1 (some people actually enjoy wallowing in themselves). Let's take the terms from Groups 2 and 3, and, as you would with lettuce, tomatoes, etc., along with ample shots of oil, vinegar, and your favorite seasonings, heap 'em together and toss 'em around in a massive salad bowl. *Voila*! What you get is an original Selfless One, pleasantly ready to rock n' roll.

Every now and then, within the eons of time, the universe decides that there's an excess of spongy people (Group 1) hanging around, helping none but themselves. And, since all global problems can easily be traced back to selfishness, Mother Earth sends some Selfless Ones (combos of Groups 2 and 3) to help lead people back onto the road of Normal Social Reality (it's located on your county map, if you look hard enough). What they *do* simply reminds us of what human society can actually *be* like if only we give out to others without worrying about what we receive in return.

The Selfless Ones are the ever handy, always helpful, never concerned about themselves humans without angelic wings species, who always seem benign to the eyes of the spongy people. And, yes, they're actually from this planet, not a nearby parallel universe.

Although they seemingly resemble everyone else (except the sponges), they differ in one respect: The Selfless Ones get a great kick out of freely helping others in whatever way they can, and do so without any concern for payment of any kind.

No, they're not so benign that they're stupid; these entities are actually quite intelligent. If Group 2 can be considered as "Oh-so smart" and Group 3 as "Oh-so pleasant," then the Selfless Ones are clever enough to realize that it's *smart to be pleasant*.

Becoming a Selfless One is not a major ordeal. It does not require a bury-yourself-alive shamanistic rite of passage. (Although, that would certainly do the trick!) All that is demanded is the stepping outside of oneself (*oh, man!*), outside your ego (*ouch*), and into the clean fresh air of self-perception (*cough*).

That constantly overworked, underpaid, mistreated, and dreadfully misrepresented human ego can now relax, take a breather, and gladly resume its original function as part of the conscious mind. This relationship allows that conscious mind to then form a natural trust with the Universe.

Unnerving, isn't it? Makes you feel rather defenseless, doesn't it? The fact is, though, that your powers and strengths have now been multiplied. Just takes getting used to, is all.

There's tremendous power in selflessness, now that the ego is back at home in the conscious mind where it belongs. A portion of that power is the freedom to openly give, without measure, whatever you can, wherever it is needed. (This is the original version of the tax write-off.)

Another portion of selfless power is the all-consuming joy one feels when offering help. It is often said that the people who understand the joy of giving also understand the joy of living, and vice versa. This tenet is so universal that it can be found in almost all (if not all) the world's cultures and religions, from antiquity on.

Being a Selfless One requires only the courage to trust that the Universe knows, and has always known, what it is doing.

During the past four hundred years, the world of science has based itself on the trust that Nature makes sense. And since all creatures, great and small, mobile and stationary, are participators in that same Universe, finding the courage to trust in Her symmetries, Her balances, should simply be a natural participatory action on our part.

The Selfless Ones are now performing a role in society that is as old as mankind itself: Helping one another solely because it is the *logical* thing to do. And this trusting relationship with the Universe just may be the key to the survival of the human species.

<><><><>

The Robin Flies Again

by Mimi Riser

<>

The shattered bones of a dead solar system?

The seeds of a new one awaiting birth?

No one knew or cared.

A navigator's nightmare was what it really was. A dense cluster of asteroids and meteorite swarms in the galaxy's center, whimsically labeled "The Forest" on most star charts. Possibly the name had begun as a joke—but the lone figure standing sentry under the spacedock bubble-dome on one of those asteroids wasn't laughing.

It might be simple coincidence he'd ended up here. Then again, it might be his own not so simple destiny. A forest was a forest, he supposed. Whether a murky tangle of trunks and brambles or a churning tumble of space debris, it could be death to the unwary or sanctuary to the desperate. Robert Loxley was the latter, and this "forest" wasn't the first he'd called home.

Home?

He sighed. Well, perhaps that was pushing things a bit. This hollowed out rock was nothing like the green land he'd once trod. There was nothing green here save the hood shading his angular features. And, of course, the expressions of those Imperial magistrates his band had recently relieved of their tax revenues. He grinned—a flash of white teeth through a close trimmed bronze beard—the grin of a wolf, or a fox.

"Poor fellows, all that wealth must have been such a burden for them. They'll feel much better out from under its weight," Loxley murmured into the chill recycled air of the dock. He'd been talking to himself a lot of late. It was the only way he could be sure of getting a sympathetic ear. His current followers were good lads and true, but...well, they couldn't be expected to understand him the way his original men had. They could hardly be termed men, if it came to that.

The feral grin waxed wistful as he gazed up through the meteorite-scarred dome. If he squinted just so, the pockmarks and scratches between him and the black velvet canopy of space made it seem almost as if he were staring at the stars through leafy branches, almost as if he were standing in a real forest of oaks and pine—the rich, fragrant forest of his birth. Was it still out there, waiting for him in some remote corner of time? When he died here someday would he awaken back on that home world of the past...as he had once died there to find himself chained in the hold of an interstellar slave trader?

God, what a rude awakening that had been. And what a stink. His expression tensed, nostrils flaring at the memory. He'd been certain at first that he'd landed in Hell. Except Hell could never have smelled so bad. That had been his second certainty. Third had come an odd sense of gratitude, although it took a while to build.

Knowing that the most accomplished slaves brought the best price, his captors had gone to extreme pains—their extremes, his pain—to broaden his education. It was largely due to them he had mastered skills he'd never have even dreamed of in his previous existence. And most graciously had he thanked them for their tutorage—right before he'd jettisoned the lot of them in a nice cozy garbage-pod. Which had been right *after* he'd learned to pilot their cruiser.

Their training of him had, perhaps, been a little too thorough? The only technology they'd avoided teaching him was their advanced weaponry. After all, one wouldn't want a slave handling anything that might prove unwholesome to his masters. What the slavers hadn't realized—unfortunately for them—was that weaponry was the one thing in which he'd needed no instruction. To an expert swordsman and archer, point-and-shoot devices like lasers, Q-beams, and magna-rays were mere child's play, almost embarrassing.

He felt the warm tingle of a blush as he unholstered his hand-mazer and set it on stun. For quite a wait now his base's scanners had been tracking a mysterious craft. A scout ship, gauging by the small size. Interesting. It hovered above him now in preparation for entry.

The dome's double hatch, operated by—*who was it today? Jock?*—at the view-screen in the base's bowels, slid silently open and then closed again. The scout floated like an autumn leaf through the pressurized airlock and settled softly on the asteroid's surface.

Poised in readiness, Loxley stood studying the docking with calm curiosity, wondering who the pilot was. These days only the very daring or very stupid ventured this deep into the Forest. Which was all the more reason to stun first and offer the introductions later. Scarcely had the first of two humanoid figures—*Herne's horns, he was a big one!*—dropped from the scout to the dock, when their welcoming committee fired his mazer.

Then gasped and jerked the beam wide of its mark.

It couldn't be!

Could it?

Build like a bear, bellow like a bull, and a wicked looking quarterstaff—*staff?*—in that ham-sized hand. It was the incongruous weapon that sealed the recognition. Loxley ducked just in time to dodge a blow from it that would have taken off his head had he been any less agile.

"History repeats itself," he groaned as a second, lower blow swept his legs out from under him. A shaggy blond bulk pinned him facedown on the dock. The *who* being certain, he was puzzling over the *how* when a voice so familiar it brought tears to his eyes growled out:

"Enough games, little man. I seek a pirate who's been raiding Imperial craft from this sector. If he's here, take me to him—quickly! Mind you, I've no quarrel with his lightening the Emperor's treasury, but I strongly resent his thieving the title of an old comrade of mine. Bodkins knows where he got the name, but 'tis said the scoundrel calls himself *Robin Hood*. I've a bone to pick with him over that, I have."

"Pick away then, you ox-brained haystack. You've found him," answered an equally familiar low timbred voice. The haystack went rigid, and the low voice added with a chuckle, "Must we enact this same farce every time you seek me, Jon? Our first meeting in Sherwood, that twig of yours toppled me into the river, I recall."

"Well, you needed a bath, didn't you, after that muck I'd tossed you in," Jonder Lyttel gruffed out. A sudden salt water wash down his cheeks gave the other man a good rinsing this time, too, as he hauled him to his feet crunched in a rib-cracking embrace. "Great Bodkins, Robin— *How*? You *were* dead, weren't you?" he blustered. "I know you were—I saw it! That nun killed you with her bloody *cure*, and I never knew what she was about till it was too late. All I could do was bury you like you asked...right where your last arrow fell..." His voice gave out, choked off by emotion.

"I appreciate the service, but apparently I recovered," Robin rasped, wondering if he'd be lucky enough to do likewise now. The only thing deadlier than Jonder's quarterstaff was his bear-hug. "As for how I landed in this time, I...I've no answer, Jon. How...how came you here yourself? How in Herne's name did a couple of sorry Saxons like us end up...in the stars?" he strangled out with his last shred of breath, and then gratefully refilled his lungs when the question instantly unlocked Jonder's massive arms.

His ruddy complexion darkened by an inexplicable blush, the big man turned away. "*Ahem*"—he cleared his throat—"I never did say I was Saxon. You simply assumed I was, and I...um, never got around to correcting you."

"What?" A pair of penetrating gray eyes widened in mock horror. "You're surely not going to confess now that you're *Norman*?"

"Not exactly." Jonder coughed again, his color deepening. "I'm a Helian."

The gray eyes went momentarily blank. "Tell me something I don't know. We were all hellions back in Sherwood. Old Notty claimed we were straight from the Devil."

Jonder's blush burned crimson. "No, I mean I'm from Helias. It's a frontier planet on the outer edge of the Beta-Gamma sector, not even on the charts till recently," he explained in an awkward rush. "To find it, you go out past Orion, veer left at the Pleiades, then straight on till you hit a constellation that looks sort of like a humpbacked quaggle picking its toes. After that, there's this tricky bit where you've got to—"

"If you follow those directions, you'll probably end up orbiting Betelgeuse," interrupted a smoky soft voice to Robin's right.

He spun about to meet it, and found himself nose to nose with a golden goddess in a one-piece suit that looked like it had been sprayed onto her.

"My navigator, Zow'ee," Jonder said wryly.

"She certainly is," Robin croaked out, sinking fast in the woman's sultry blue eyes.

An even deeper blue gaze yanked him out of it—*damn her*. Silently, he cursed the elfin figure hovering wraith-like before his inner vision. *Can't I even look at another maid?* The beguiling little witch gave him no peace, did she? Marian's memory was more distracting than the flesh and blood version had been. She was always—*always*—there, hanging 'round his heart like a millstone, grinding away at him without mercy.

"A constant headache is what she is," Jonder grumbled with a glowering glance at Zow'ee. "Good in a fight, I'll grant you, but laboring under the misguided delusion I can't find my own arse without her calculations."

"Truth hurts," Zow'ee purred. Her eyes slanted from one man to the other before focusing their full attack back on Robin. "Jondy has many marvelous talents, but astro-trig isn't one of them," she said, somehow making the comment sound like a deliciously indecent proposition.

It was a wasted effort by that point, but its recipient appreciated the thought behind it, nonetheless. Still struggling with Marian's haunting image, he plastered what he hoped was a reasonable approximation of polite attention on his face while Zow'ee added:

"That's how Jondy ended up in your era in the first place. He'd been heading for a rebel base near Alpha Centuri, but he overshot—of course—because he'd left me behind. When he didn't report in as scheduled, I'm the one who finally located him with a spacetime-tracker and rescued him."

"Aye, lucky me," Jonder groused. "And it was hardly an *overshoot*. I was only about four light years off."

"Four light years *and* five millennium," Zow'ee amended. "Your biggest mistake was using that wormhole."

"I was in a rush, thought it would save time. Only it saved a bit too much. Well, how was I to know the blasted worm was still in it?" Jonder defended himself. "Wasn't one of your generic cross-D worms, neither. Turned out to be a *temporal*—a *hungry* temporal. Sucked me right down its gullet, absorbed my power packs, and when I slid out the tail, there I was in a drained ship right over twelfth-century England of all places. Crash-landed in a pigsty just outside Nottingham. Scared hell out of the old couple who owned it. They didn't know what to make of me. Spent all night arguing over whether I was an angel or a devil."

"A little of both, I'd say," Robin cut short the confession. His tight expression relaxed a notch. "Now we'd best find some warmth. It's damned cold out here." Turning to lead the way off the dock, he inquired almost too jovially, "How long has it been since your last meal?"

"Thanks, I thought you'd never ask," Jonder replied as the three crowded into the turbo-chute that would whiz them below the asteroid's stark surface to the control rooms and living quarters in its belly. Though, truth be told, he hadn't much appetite at the moment. He'd been given too much to chew on already. *Robin? Here? Now?* Bloody Bodkins, he'd never digest this turn of events. He didn't even want to try. Far easier to just blame in on a wicked whim of Fate. 'Twas often said the Gods had a warped sense of humor.

* * *

I could crisp Jondy for setting me up like that. Zow'ee silently seethed. *Him and his brilliant ideas. "Once we find this renegade, I'll knock him off balance, then you keep him that way," he told me. "Turn on the charm, m'girl. Whoever the liar is we want his cooperation, don't we? The Liberty Alliance needs a safe passage through this sector."*

In actuality, the Liberty Alliance needed organization even more—and a strong, sure hand at the helm most of all. The rebel worlds all craved freedom, but rarely agreed on how best to achieve it. They spent as much time arguing among themselves in secret council chambers as they did fighting Imperial oppression. But Zow'ee wasn't thinking of that right then.

"Charm, my Aunt Fanny. What did I do wrong?" she muttered under her breath as the chute's transport-pod jolted to a stop. Rejection was not a common experience for her. In fact, it was a completely new experience—the first time ever a humanoid male hadn't melted into a quivering glob of malleable slag under her skillfully applied heat. And first experiences are often unsettling, even when they're pleasant. Which this one was not. Allowing herself the indulgence of a small but lethally provocative pout, she preceded the cause of her upset into the base's communal hall, swishing her perfect Aunt Fanny before him just for spite.

A few swishes into the room she stopped short and stared. *Gag!* If this was the Acme Astro-Corps's idea of soothing surroundings, it was small wonder they'd gone bankrupt. The large chamber was a disconcerting ellipsoid shape—all glaring pinks and chrome—studded with tri-D screens, and dizzily swarming with free-floating 'orm chairs and bobbing glo-globes, remnants from its days as a resting and refueling stop for space-sick traders.

And if they weren't sick when they arrived, they must have been by the time they left.

At the far end, clashing with the décor, sat a stocky, bright blue dwarf with the bristling green topknot and double thumbs of a Changolian.

Barbarian race. *Ick.* She hoped he was housebroken. He was playing a high-speed round of Jabberwok with a rusty, rotund robo. One of those old TUK models, probably. Which junk-world did they salvage him from? And *why*, she was wondering irritably when a nearby wall panel slashed open and the unmistakable racket of Galusii swine being slaughtered split through the hall.

"Hoot mon!" a voice bellowed as the squealing slid to a teeth-grinding halt. It came from a tall human in a preposterous pompom-topped cap and short plaid skirt. "On spying yon bonny, braw lassie throo the dock monitoor, methought tae welcum her wi' a tune on me pipes!" he brogued, tossing the instrument onto a passing 'orm and dipping a low bow before Zow'ee. "Jock of Aberdeen at yoor service." He grinned.

"*Today* he's Jock of Aberdeen." Robin sighed and pushed back his hood with a long-suffering sweep of his hand. "Yesterday it was Chief Eagle Feather followed by Erik the Viking. And the day before that he spent romping about in green tights, shooting arrows at the 'tyrannical' glo-globes and distributing their ill-gotten wealth to the poor peasant 'orm chairs."

"Ah, 'tis just jealous he is that me own Robin o' Sherwood puts his tae shame." Jock chuckled and shot Zow'ee a rakish wink. "By the way, I've erased that program fray me mem'ry banks," he informed Robin.

"Very kind of you, I'm sure," came the dry response.

"'Twasn't kindness. Those tights were givin' me the itch, they were." Jock said, beaming his grin back to the bonny lassie.

Her mouth, which had dropped open at his entrance, snapped shut into a hard, disapproving line.

"What? Ye dinna care foor Highlanders?" he asked. "Then purhops ye'd prefer this!" Instantly he began to shimmer about the edges. The plaid mini-skirt lengthened and split into some primitive kind of coarse, faded blue pants, while the pompom cap transformed into a big, broad-brimmed head covering that looked as though it could double as a snerx pail.

"Pleased to make yer acquain'ance, ma'am," its wearer drawled. "Tex is the name. Ridin', ropin' and brandin's my game."

"Great Bodkins!" Jonder spouted. "He's a hologram."

"Am not! I'm solid as you are, y'overgrown buff'lo." Tex thumped his sturdy cowboy chest.

Zow'ee quickly stepped back a repulsed pace. "Oh, gross, he's an *android*."

"Y'all gots sumthin' 'gainst androids, sweet pea?" Unperturbed, Tex reached forward and chucked her on the cheek.

She answered with a backhand blow to his jaw that nearly spun his head around.

"*Whoo-ee*! Feisty little filly, ain't ya? I do like a gal with spirit," he chortled, happily basking in the attention.

"He's a holoid," Robin explained, before the *attention* got out of hand. "We found him in stasis on an antique capsule our tractor beam snagged. I activated him out of curiosity"—he winced—"and haven't been able to turn him off since."

"Not 'a' holoid," Tex fussed. "I keep tellin' y'all, I'm *the* Holoid—only one ever made. I was created fer the ATTS Project. That stands fer *Ambassadors to the Stars.* I'm a walkin', talkin' introduction to humanity!" he stated proudly.

"According to the data we discovered in his capsule, the ATTS project took place on Earth during the twenty-second century," Robin elaborated. "It's purpose, apparently, was to send hundreds of contact-probes out into the galaxy, each one containing a holoid programmed to perform several dozen persona spanning the gamut of terran man's history and cultures."

"But the funds fizzled out right after me," Tex added sadly.

"Pity they couldn't have fizzled out just a little sooner," Zow'ee said, her gaze hitting him like a freeze-beam.

Jonder studied the enigma with narrow-eyed interest. "What exactly *is* a holoid?"

"An android with a specialized holographic interface that enables him to modify his appearance," Robin replied. "Only within certain parameters, though. Primarily just his costume, thank Herne. He's difficult enough as it is. It would be impossible to manage him if he could vary his size or sex."

"Sex?" The holoid's face lit up like a nova. "Now you're talkin'!" He advanced wolfishly on Zow'ee. "Hey, honey babe, you ever make it with an artificial intelligence?"

Zow'ee turned green. "I think I'm going to be sick."

"Shucks, that's just nervous excitement. C'mon, sugar, don't be shy," Tex persisted. "I got a Casanova program that'll blow yer mind."

"Not if I blow yours first," she said, bringing her leg up sharply. A slash of light knifed out from her kneecap.

Tex dove for the deck, and the glo-globe that floated into the position his head had been occupying exploded with a loud sizzling pop.

"Dang!" A second flash from her elbow drove him rolling to the left. "If this is her idea of foreplay, I'm gonna need sumthin' in armor. Whoa, baby doll, gimme a sec to put on Bert, will ya?" he pleaded while dodging behind a large freestanding tri-D screen to avoid the third and fourth shots from her shoulder and hip.

"A micro-mazer studded suit?" Robin wondered aloud, watching the activity with semi-detached professional curiosity.

"A micro-mazer studded body." Jonder lunged forward and hoisted Zow'ee off her feet with a burly arm about her waist. Before she could grapple loose, his other hand found the switch under the thick coil of hair at the nape of her neck, and she collapsed against him, limp as a Pogling marionette whose strings had been cut. "She's a bionic. Woman's brain, but mostly synthetic form. Overly sensitive about it, I'm afraid—hates androids because they remind her of her own semi-human status. I'll turn her back on after her circuits have cooled." Gently, he deposited her into a waiting 'orm chair.

"If only I could deal with you so easily," Robin said to the dusky armored figure clanking cautiously out from behind the tri-D screen. "Is it too much to hope that you've been damaged?"

"Alas and alack! 'Tis most grievously wounded I am—though not in body. Methinks I am sore in love with yonder positronic-powered damsel," Sir Ethelbert the Black Knight wailed. "To win her favor, I do herewith dedicate my sword and my lance to do her homage. Squire, my steed! Let the joust begin!" The crimson plume in his helmet quivered with passion as he brandished his broadsword aloft.

"Don't look at me. I haven't been near a horse in fifty centuries," Robin told him with a dry smile. "Why don't you try Lao Tse Ching-Ting, instead? Perhaps a holy man would capture her interest."

Sir Ethelbert wavered in doubt a moment; then decided, apparently, he had nothing to lose. "A noble suggestion!" he proclaimed with a flourish of his sword that sent glo-globes scurrying in all directions. "By your leave, my liege—"

He crashed backward into the path of the nearest 'orm. It dipped to catch him, molding close to his ironclad contours, which had already begun to shimmer and soften. By the time the armor had metamorphosed into the saffron robes of a Buddhist monk, Lao Tse was curled lotus-like in the hovering seat, peacefully chanting *Om*.

"If we're lucky, he'll be busy trying to achieve *nirvana* until his next shift at the monitors," Robin said. Closing his eyes, he sank against a second 'orm. It flattened against his back, lengthening to support him in an upright lounge.

Jonder settled into a third one, which heroically strained its dimensions to accommodate his bulk. "What about you?" he asked, careful to keep his tone casual.

"Me? I'm not even sure what nirvana *is*, let alone how to achieve it. I only learned the term from his 'holiness' here. But he doesn't seem to be entirely sure what it is, either. Do you have any idea wh—"

Jonder cut him off with a half-hearted chuckle. "Don't be cute. You know what I meant. You're taking this very calmly, Rob—even for you—the news I'm a Helian and all. Not even asked why I'm here. I came in an unmarked craft, for godssake—in the thick of a very dirty war where nothing and no one is safe! How can you be sure I'm not an Imperial spy, or a rogue playing both sides of the battle?"

The hooded eyes opened into a guileless gray gaze. "A rogue, for certain. A dishonorable one? Never."

Jonder found his own eyes suddenly moist. "You trust me that much, do you?"

"With my life, Jon, as I always have." A weary flick of the hand brushed the matter aside. "Anyway, I stopped asking *why* the day I died in Sherwood and awoke in the middle of a million worlds under siege. I simply take what comes and deal with it as best I can."

"That's what you've always done. You're a born opportunist," Jonder said, his chuckle genuine this time.

"You're probably right about that." The figure opposite him chuckled in return. "It does get a bit tedious, though. In five thousand years we haven't learned a damn thing that I can see. The rich get richer and bleed the rest of creation dry. Nothing ever changes, does it, Jon? The tyranny never stops."

Pulling away from his backrest, he paced the hall, a man in obvious need of a rest, but knowing there was none to be found, not for him. "This is Sherwood all over again, isn't it? The same fight we had there. Only bigger."

"Bigger? It's galactic!" Jonder let out a harsh laugh. It ended short as his breath caught in his throat at a suddenly rekindled awareness.

"Galactic, but...um, maybe not impossible," he added awkwardly. His heart began to race with something he hadn't felt in far too long. Something that could almost be called...

Hope?

Like a hungry hawk, his gaze followed the pacing figure.

Blast me, I'd forgotten his power, why we won so often back in Sherwood. It was madness what we did. We should have been stomped into dust. Just a tiny, ragged band we were—against a royal army! But you'd never have guessed that. Why? Because he made us feel that we *were the kings.*

That was Robin's secret, Jonder decided. Not his courage or his cleverness or his strength, but his incorrigible confidence. And his ability to inspire the same in others. His indomitable will was a magnet, pulling the best and better than the best out of all he touched. When the Hooded Man led, hopeless fights had a way of flipping head over heels into victory.

Bodkins knew such a leader was needed now more than ever.

"Galactic, but not impossible," Robin repeated with the ghost of a grin. "Thank you, Jon. Your childlike faith in the ultimate triumph of justice over oppression always did help me keep things in perspective."

Halting his pacing by the Jabberwok players, his grin broadened. "And if you two don't put away those weapons, I'm going to break that game screen over your heads," he pleasantly informed the dwarf and robo. His gaze shifted back to the giant in the laboring 'orm. "Jon, you have no idea how close you came to becoming a Changolian shish-kabob. Mooji's skill with a laze-blade is the stuff nightmares are made of. And Tuk's polybeams are almost as horrible."

The blue dwarf blushed purple, and the robo emitted an electric crackle that sounded suspiciously like embarrassed chuckling.

"Can't you tell a friend from a foe?" Robin asked them.

"Better safe than sorry," Mooji grumbled. He looked unconvinced, but with a flick of his second thumb, he deactivated his blade and slipped it into a fold of his tunic.

"The large one will be joining us?" Tuk's metallic voice buzzed out.

"I don't know," the large one answered. His gaze locked on Robin across the room. "I haven't been invited yet."

"I didn't realize I needed to," Robin said quietly, holding the stare unblinking while Jonder blushed darker than the dwarf. "You're your own man, you know, free to go or stay as you please."

What a wonderfully crystalline statement. It clarified things for Jonder to a fine, sharp tee. The breath he'd been holding hissed out in a raspy curse.

"Bogrop. That choice was taken from me long ago, and we both know it," he said, surrendering to warm drowning in those bottomless gray pools. "I'm *your* man, Robin-in-the-Hood. As I once promised in Sherwood, I'll follow you to the grave and beyond."

The gray gaze crinkled into wry devilment. "I remember. And it appears that's exactly what you've done."

"Aye, that I have," Jonder had to agree. Laughter bubbled up inside him at Fate's twisted humor.

Then again...

In his mind he saw many others soon following a legend's lead. All those dedicated but disorganized rebels finally united into one solid successful force?

Why not? Far stranger things had happened—as he could well attest.

Well, I'll be damned.

Maybe it hadn't been a cosmic joke, after all, that brought Sherwood's *rebel-rouser* here. All things considered, maybe Robin was exactly where and when he was supposed to be.

==============

<><><>

About the authors:

Rob Riser spent twenty-five years as a professional musician, playing everything from late '60s Rock to big-city Symphonic. He now spends his time pursuing the mysteries of mathematics and physics, and analyzing a lifetime of paranormal experiences.

Mimi Riser is a best-selling, award-winning author of fiction and nonfiction. She claims that living with Rob *is* a paranormal experience.

Both began life in the urban northeast, but now reside in the rural southwest with two dogs and more cats than anyone wants to know about.

http://www.mimiriser.com

==============

INTERVIEW WITH ROB RISER:

<> Rob, what do you think about modern science?

R: Are you asking what I think about what modern science *creates*, what I think about how it *uses* what it creates, or whether or not I think Modern Science *thinks* at all?

<> Er...let's take...ah...

R: Well, thinking is thinking is thinking. Thinkers know this. They enjoy collecting concepts and solving problems. And there is always so *much* to think about that if you enjoy it, thinking goes on forever. Isn't that great?

<> Um, I think I need a beer. Thanks, Rob. No more questions.

==============

OTHER BOOKS BY MIMI & ROB:

<>

Wonderful Weird True Tales
The Kitchen Witch Spring into Summer Book
The Kitchen Witch Yuletide Book

OTHER BOOKS BY MIMI RISER:

<>

Novels

Eyes of the Cat

Every Jack Needs His Jil

Sherwood Charade

Space Rats and Rebels

Short Stories

If Wishes Were Kisses

The Great Race

Joy to the Worlds

Nonfiction

The Kitchen Witch Glossary of Cooking Herbs & Spices

The Kitchen Witch Guide to the Timeless Art of Herbal Tea

The Kitchen Witch Glossary of Herbs for Magical Use

Spice Up Your Love Life! Aphrodisiacs from the Kitchen

The Naughty Girl's Guide to Divining by the Dots

The Kitchen Witch Halloween Book

Erotic Romance

Cymric's Rose

Samantha White & The Seven Dwarves

Dungeons & Dirty Dreams

Romeo's Revenge

Tina Takes a Tumble

Saving Sally Savoy

A Knight to Remember

Just Another Lonely Knight

In the Heat of the Knight

The Adventures of Cassie Nova

The Cowboys & The Courtesan

Wicked Comes the Beast

Playing Pirates

Pirates Do It With Passion

The Sylver & Steele Series

Anthology

Romance: Fantasy, Food, and Fortune Telling

Paperback Anthologies

My Knightly Adventures

Return to the Burn

Can't Fight the Feeling

Pirates & Other Wicked Pleasures

=======

<><>

<>

www.ingramcontent.com/pod-product-compliance
Lightning Source LLC
LaVergne TN
LVHW090335210926
848480LV00004B/1426